Brilliant Support Activities

Understanding Living Things

Janet O'Neill, Alan Jones and Roy Purnell

Brilliant Publications

We hope you and your class enjoy using this book. Other books in the series include:

Understanding Physical Processes	1 897675 61 5
Understanding Materials	1 897675 60 7

Published by Brilliant Publications,
The Old School Yard, Leighton Road,
Northall, Dunstable, Bedfordshire LU6 2HA

Written by Janet O'Neill, Alan Jones and Roy Purnell

Designed and illustrated by Small World Design

The authors are grateful to the staff and pupils of
Gellideg Junior School, Merthyr Tydfil for their help.

Printed in Spain by E.G. Zure. 48950 - Erandio (Bizkaia)

First Published in 2000
ISBN 1 897675 59 3

Contents

	Link to NC	*page*
Introduction to the series		4
Introduction to the book		5
Human development quiz	1b, 2f	6
Growing up	1a, 1b	7
Bones, bones, bones	2e	8
Skeletons	2e	9
Muscles	2f	10
Teeth	2a	11
Food for life	2b	12
Good food	2b, 2g	13
Pulse	2c, 2d	14
Growing plants, 1	3b	15
Growing plants, 2	3b,3c	16
The heart of the matter	2c, 2d	17
Heart	2c, 2d, 2g	18
Blood	2c	19
The body	2c	20
Exercise	2c, 2d	21
Healthy living	2g	22
Staying healthy	2g	23
Unhealthy living	2g	24
Words for people	2g	25
Living and non-living	1a, 1b, 1c	26
Alive and not alive	1a, 1b	27
Living things	1a, 1b	28
Dying plants	1b	29
Seeds, 1	3d	30
Plants in the cold	3a	31
Pollination	3d	32
Seeds, 2	3d	33
Different plants	5a	34
Guessing animals	4a, 4b	35
What animal?	4a	36
Making keys	4a	37
Changes in a tree	3d	38
Frog cycle	5b	39
Mini-beast families	4a, 4b	40
Environment	5a	41
Where do I live?	5b	42
Plants live here	5b	43
Where fish live	5b, 5c	44
Food chains	5c, 5d	45
Some animals eat other animals	5c, 5d	46
Diseases	5e	47
Micro-organisms	5e	48

Introduction to the series

This series is designed to help the slower learner or pupils with learning difficulties at Key Stage 1 and 2 develop the essential skills of observation, predicting, recording and drawing conclusions. These pupils often have been neglected in more conventional commercial schemes of work. The books contain a mixture of paper-based tasks and 'hands on' activities. Symbols have been used to indicate different types of activities:

What to do

Think and do

Read

Investigate

The sheets support the attainment targets for Key Stage 2 Science, Section 1: Scientific enquiry and Section 2: Life processes and living things. The practical investigations use materials readily available in most primary schools. The activities have been vetted for safety, but, as with any classroom-based activity, it is the responsibility of the class teacher to do a risk assessment with her/his own pupils in mind.

The sheets usually introduce one concept or National Curriculum statement per sheet (unless they are review sheets). The sheets are designed for use by individual pupils or as a class activity if all the class are working within the same ability range. They can be used in any order, so that you can choose the sheet that best matches a pupil's needs at that particular time. As with any published activities, the sheets can be modified for use by specific pupils or groups. The sheets can be used as a support for your present schemes, as an assessment task, or even as a homework task. If used for assessment purposes, then you will need to devise a marking scheme or level indicator. Generally the sheets are designed for use with levels 1–3 but some can be used at level 4.

The sheets use simple language and clear, black line illustrations to make them easy to read and understand. They have been tested to check that they can be understood by pupils with learning difficulties. Although the sheets have a reduced vocabulary, they encourage pupils to produce written responses and to develop their writing skills.

No particular reference has been made to any type of disability as the activities should be accessible to a wide range of pupils and it is up to the teacher to select the most suitable modes of access to match the needs of his/her pupils. For example, the activities could be photo-enlarged, converted to raised tactile diagrams, or recorded on an audio tape.

Introduction to the book

The topics in this book help pupils understand life processes and living things. They reinforce methods of scientific enquiry by requiring pupils to plan and carry out practical activities, consider evidence, and present ideas and conclusions. The sheets focus on life processes common to all living things, humans and other animals, green plants, as well as on how living things relate to their environment.

The worksheets in this book overlap and you will find that several statements of the National Curriculum are covered several times in a number of different ways. This is to allow you to use the worksheets to repeat work on particular concepts to reinforce your pupils' learning. However, the worksheets are not designed to be used in any particular sequence. They are not a teaching scheme, but are a resource which you can use to enrich or augment your own particular scheme of work according to the needs of your pupils.

Some worksheets encourage an open-ended response, others are designed to lead pupils to a particular answer. Some start with easy tasks and progress to more difficult extension activities which we have called 'Think and do'. Others are at one level of difficulty. The variety is designed to give the worksheets flexibility and to allow you to select the most appropriate worksheet for your pupils.

Human development quiz

What to do

Search out these words to do with people.

baby	boy	girl
adult	you	old
teen	age	grow

b	w	t	z	g	i	r	l
a	d	u	l	t	x	e	n
b	b	o	y	b	o	l	d
y	o	u	l	r	a	g	e
t	e	e	n	w	o	r	g

Think and do

How many words can you make out of the word:

teenager

Hint: it is good if you get 10 words at your age!

Growing up

What to do

As we get old our bodies change.
Put the words next to the pictures:

baby **teenager** **girl** **toddler**

boy **adult** **old age**

b _ _ _

t _ _ _ _ _ _ _

g _ _ _

t _ _ _ _ _ _

b _ _

o _ _ _ _ _

a _ _ _ _

Why are babies smaller than toddlers?

Think and do

What is the youngest you can be a teenager? __________ years

How old is the oldest teenager? __________ years

At what age do men get state pensions? __________ years

Bones, bones, bones

What to do

Sort the letters to find the parts of the body.
Draw lines from the words to the correct parts of the body.

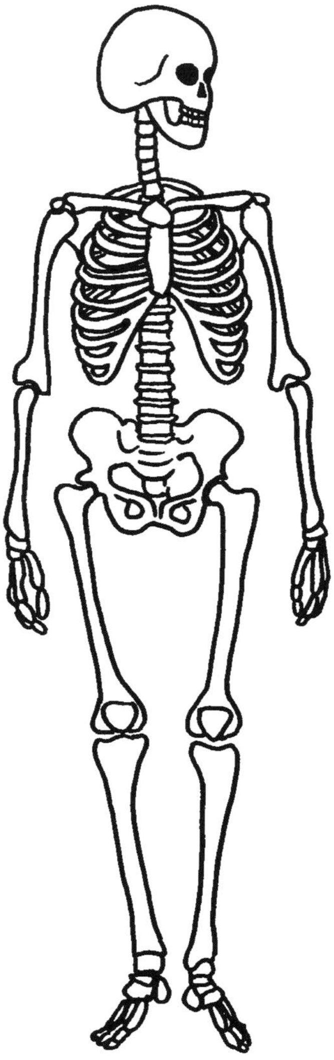

The bone of the head
is a **sullk.** . . . .

The bones of the chest
are **birs.** . . . .

The bone that holds the body
up is the **cabk** bone. . . . .

The leg bends at the
keen. . . . .

Label any other bones you know.

Think and do

Fill in the missing words:

Our brain is inside the **s** . . . .

The skeleton is made of **b** . . . .

Colour in the **pelvis** on the picture.

Skeletons

What to do

Put a tick ✓ by the picture if the animal has a skeleton.

Kangaroo

Skeleton Yes ☐ No ☐

Slug

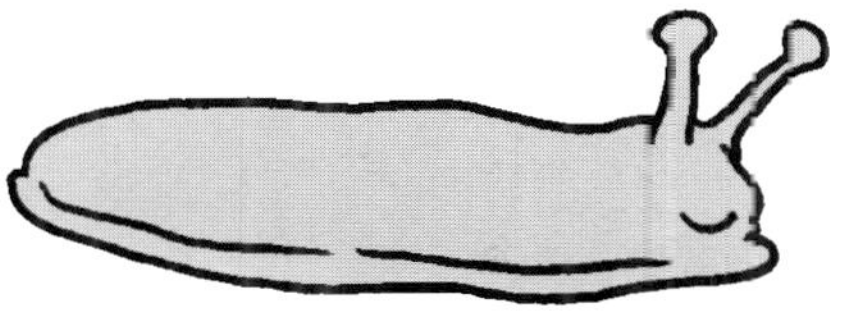

Skeleton Yes ☐ No ☐

Baby

Skeleton Yes ☐ No ☐

Bird

Skeleton Yes ☐ No ☐

Fish

Skeleton Yes ☐ No ☐

Frog

Skeleton Yes ☐ No ☐

Think and do

Why do we have skeletons?

. . . .

Do jelly fish have skeletons? . . . .

Muscles

What to do

Fred wants to be able to pick up heavy weights.

What does he need to build up?

Cross out the incorrect words.

brain/muscles/bones

Jo wants to throw the ball further.
Which muscles does she need to strengthen?

Which muscles does Ranjit use when he kicks the ball?

After exercising a lot, what aches?

Cross out the incorrect words.

bones/blood/muscles

Think and do

List the muscles you use most during the day.

. . . .

What happens to muscles if you don't use them?

. . . .

Teeth

What to do

Look at this picture of teeth in your mouth.

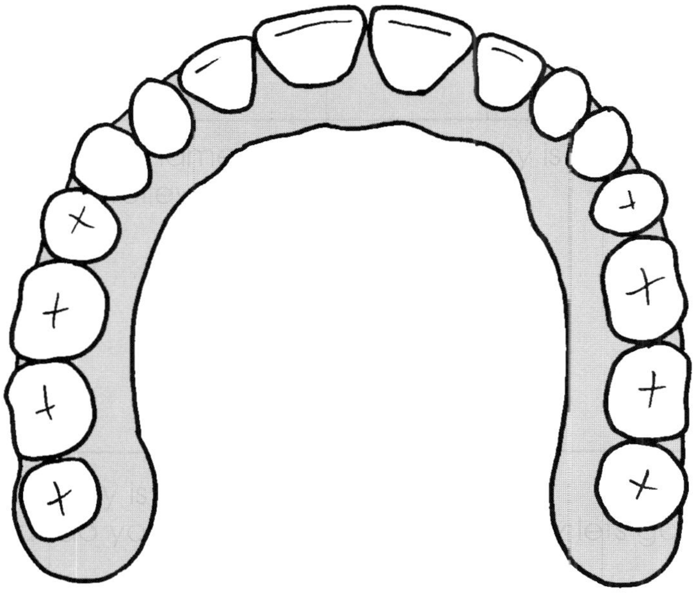

i _ _ _ _ _ _

c _ _ _ _ _

m _ _ _ _

We have three types of teeth: molars
incisors
canines

Draw lines from the labels to the tooth picture.

Think and do

Which teeth do we use for biting our food? ....

Which teeth do we use for grinding our food? ....

What do the other teeth do? ....

Food for life

What to do

Work out the missing words.

energy
healthy
grow

... give me

e _ _ _ _ _

... help me

stay alive and

g _ _ _

... keep me

h _ _ _ _ _ _

A car needs petrol fuel to keep going.

F _ _ _ is like the fuel to keep me going.

Think and do

My favourite food is ...

It is important to have breakfast before coming to

school because . . . .

. . . .

Good food

Read

You must have a balanced diet of good food to keep healthy.

What to do

Jo and Di had lunch boxes.

crisps
cream cake
chocolate bar
can of coke

apple
cheese
brown bread roll
and salad filling
carton of milk

Who had a well-balanced meal? . . . .

Think and do

Write down what you had for lunch.

Was it a well balanced meal? Yes ☐ No ☐

Why? . . . .

. . . .

Pulse

Read

The heart pumps blood around the body.
It beats about 72 times a minute.

What to do

Work out the missing words:

As the heart pumps, you can hear your

h __ __ __ __ **b** __ __ __ . You can also feel your pulse.

A doctor uses a **st** __ __ __ __ **scope** to listen to your heart.

Investigate

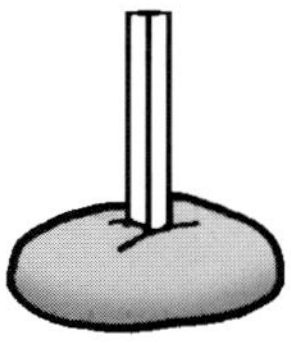

Make a shape like this with some Blu-tack or Plasticine and a matchstick.
Put your arm palm upwards on a table.
Put the Blu-tack on the pulse point on your wrist or on the side of your thumb.

Write or draw what happens.

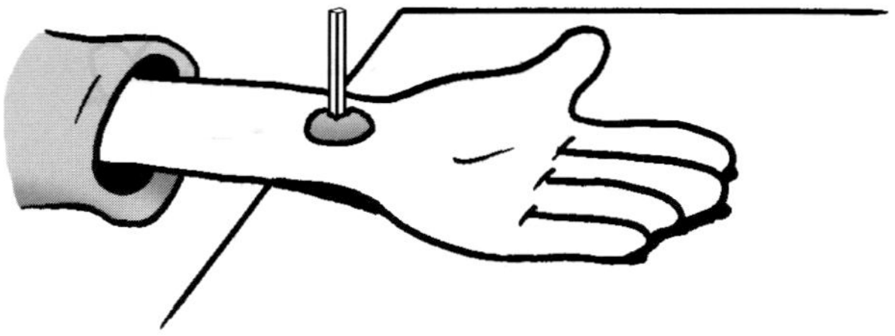

Count how many pulses in one minute: . . . .

Think and do

Rob ran around at play time.
Does his pulse rate go up or down?

You could investigate this.

Growing plants, 1

What to do

Paul has some tomato seeds. He wants to grow tomato plants.
He wants to sell them to make some money.

Use these words to complete the sentences:

seeds	pot	soil
sunny place	water	label

1. Paul put the **s** __ __ __ in the **p** __ __ .

2. Then he added the **s** __ __ __ __.

3. He added **w** __ __ __ __.

4. He put a **l** __ __ __ __ on the pot.

5. Then he placed the pot in a **s** __ __ __ __ **p** __ __ __ __ .

Think and do

What would happen if the seeds were not watered often? . . . .

What would happen if the pot was put in a dark place? . . . .

You could investigate this.

Growing plants, 2

What to do

Name the parts of the plant.

Draw lines from the names to the plant.

leaves	flower	roots	stem

f . . . .

l . . . .

s . . . .

r . . . .

Work out the missing words:

Plants need water to live.

Plants take up water from the soil using their **r** _ _ _ _.

The stem carries the water from the roots to the **l** _ _ _ _ _.

Think and do

Can a plant grow well in the dark? Yes ❑ No ❑

What does the flower attract?

Sort out the letters: **i c t s e n s** i _ _ _ _ _ _

Why do some plants attract bees?. . . . .

. . . .

The heart of the matter

What to do

Draw an arrow to show where your heart is in your body.

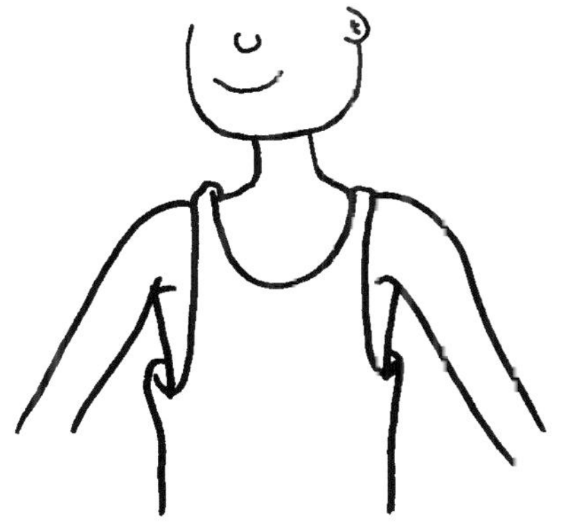

How big is your heart?
Put a tick ✓ by the correct answer:

My heart is the size of a

- pea. ❑
- fist. ❑
- football. ❑

Label the diagram to show the parts that let the blood go around in the body.

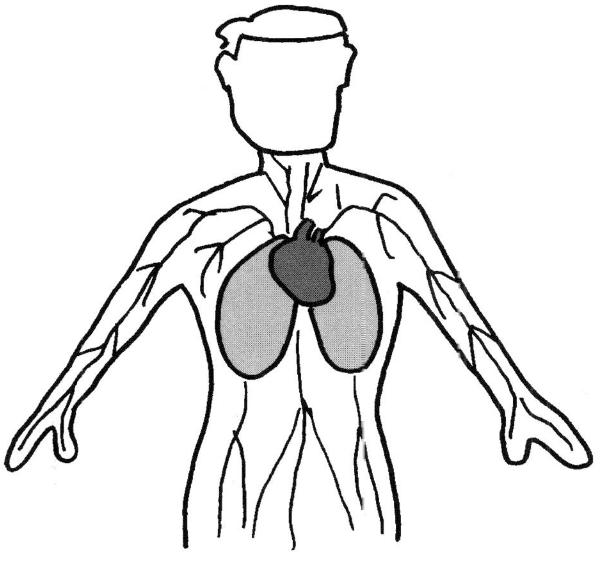

Veins

Arteries

Heart

Lungs

Think and do

What happens to your heart when you run?

. . . .

What happens to your heart when you go to sleep?

. . . .

Heart

What to do

Put a tick ✓ by the sentences that are correct.

☐ The heart pumps blood around the body.

☐ I feel love and hate with my heart.

☐ The heart slows down when you sleep.

☐ Smoking can give you heart trouble.

☐ Your heartbeat gets faster when you run.

☐ The heart only starts to beat when you are born.

Think and do

Do animals like cats and dogs have hearts? Yes ☐ No ☐

Do plants have hearts? Yes ☐ No ☐

Blood

What to do

Put a tick ✓ in the correct box.

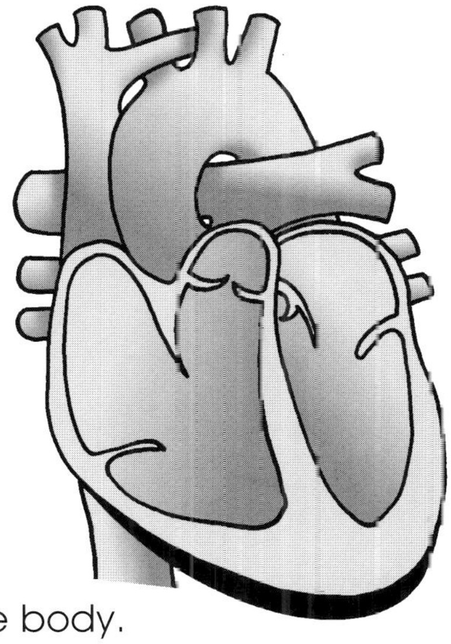

Blood is
- red. ❑
- green. ❑
- yellow. ❑

Blood carries
- chocolate ❑ around the body.
- oxygen ❑
- juice ❑

Blood is pumped by the
- brain. ❑
- lungs. ❑
- heart. ❑

There are about
- 2 ❑ pints of blood in an adult body.
- 7 ❑
- 10 ❑

Think and do

Do animals like cats and dogs have blood? Yes ❑ No ❑

Do plants have blood? Yes ❑ No ❑

The body

What to do

We have things called **organs** inside our bodies.

They are the:

brain	heart	kidneys	lungs

Draw lines to the body.

l_ _ _ _

k _ _ _ _ _ _

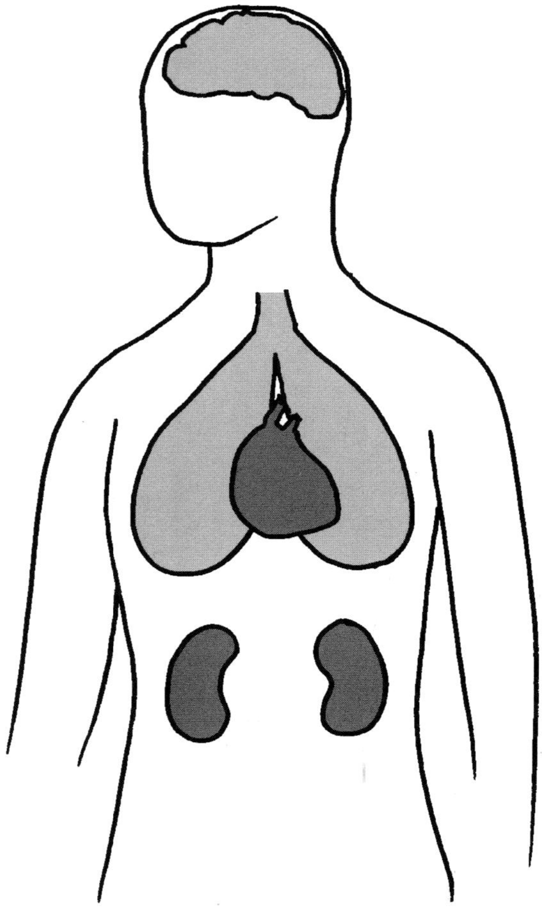

b_ _ _ _

h _ _ _ _

Think and do

Which organ pumps blood around the body? . . . .

Which organ can take in air? . . . .

Exercise

What to do

Put a tick ✓ by those things that change when you exercise (say, run fast).

How do they change when you exercise?

	Change? Yes/No	Say how they change
Breathing		
Blood moving around body		
Heartbeat		
Body temperature		
Your teeth		
Your fitness		
Hair growth		
How tired you are		
Sweating		

Think and do

Why is it good to exercise on a regular basis?

. . . .

Healthy living

What to do

Put a tick ✓ by those things that help good health.

Smoking ☐

Drinking alcohol ☐

Regular exercise ☐

Balanced diet ☐

Eating fruit ☐

Sniffing glue ☐

Eating sweets ☐

Peaceful life ☐

Think and do

Why is smoking bad for you? . . . .

What happens when you become unwell? . . . .

Staying healthy

What to do

We wash our hands, to help stop stomach upsets.

Put these phrases in the correct places.

Eating less sweets	Staying away from cigarette smoke
Coughing into a handkerchief	Regular exercise

. . . . helps to stop the spread of colds.

. . . . helps to stop toothache.

. . . . helps to keeps you in good health.

. . . . helps you to stop getting lung cancer.

Think and do

Write three things you do to stay healthy.

1. . . . .
2. . . . .
3. . . . .

Unhealthy living

What to do

Put a cross ✖ by those things that are unhealthy.

- Smoking ☐
- Balanced diet ☐
- Alcohol ☐
- Over eating ☐
- Sniffing glue ☐
- Polluted air ☐
- Pure water ☐
- Medicines ☐
- Drug abuse ☐

Draw a poster for healthy living.

Think and do

I keep healthy by ...

. . . .

. . . .

Words for people

What to do

Draw lines to match the words on the left with the phrases on the right.

Doctors	**treat animals.**
Physiotherapists	**give out medicines.**
Dentists	**look at your feet.**
Vets	**treat people.**
Chiropodists	**exercise your body.**
Pharmacists	**treat your teeth.**

Think and do

What do they call a person who takes a pet stone for a walk?

A fool!

Why do children love to play with 'electronic pets' such as Tamagotchis?

....

Living and non-living

What to do

Put a tick ✓ by the things that are alive now.

Tree ☐

Wooden pencil ☐

Coal ☐

Corn flakes ☐

Flower ☐

Bird ☐

Mouse ☐

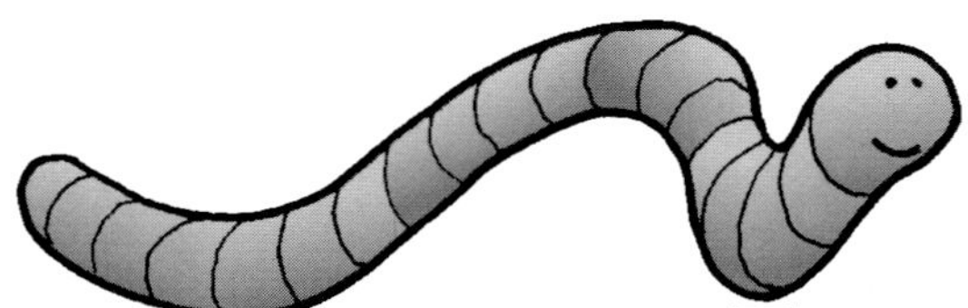

Worm ☐

Think and do

Draw a ring round the things that all living things need.

Water **Food** **Holidays** **Air** **Light**

Alive and not alive

What to do

Some of the things below are alive.
Some are not alive.

Grass

Water

Flower

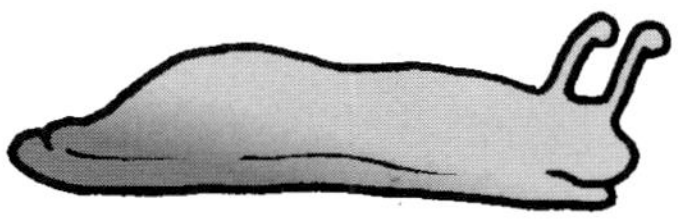

Slug

Cloud

Crab

Rain

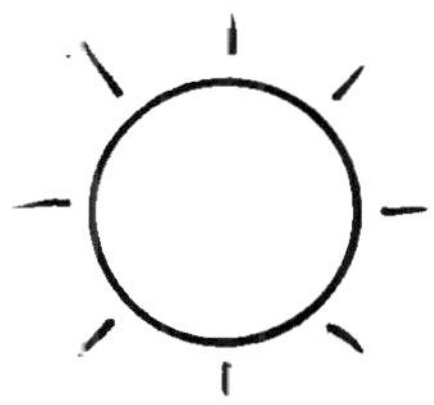

The Sun

Toothbrush

Put the words in the correct list:

Alive	Not alive
....	....
....	....
....	....
....	....
....	....
....	....
....	....

Living things

What to do

Put a tick ✓ by things that all living things do.

Need food ❑

Reproduce ❑

Need water ❑

Like music ❑

Need light ❑

Grow ❑

Move ❑

Watch television ❑

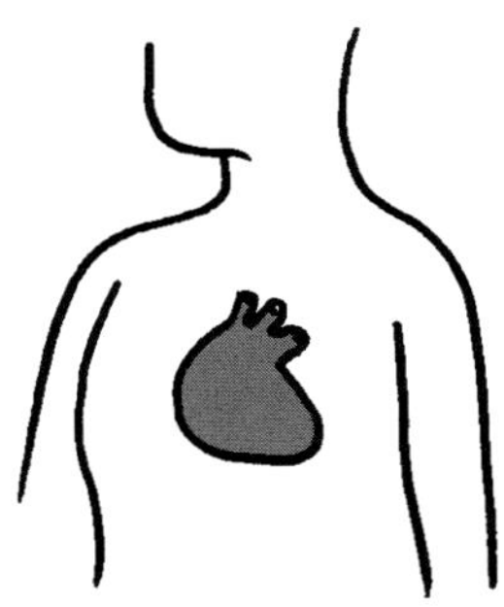

Have a heart ❑

Think and do

Animals have blood which takes the food around the body.

What do plants have that does this? ....

Dying plants

What to do

Here are some plants.

1. Water
   No plant food
   Sunlight

2. No water
   Plant food
   Sunlight

3. Water
   Plant food
   Grown in the dark

4. Water
   Plant food
   Sunlight

Which do you think will grow the best? ....

Why will the rest not grow so well? ....

....

You could investigate this.

Think and do

A plant will not grow on the moon's surface because ...

....

....

Seeds, 1

What to do

Some plants reproduce by making seeds.

Draw how a tomato plant grows from seed.

1. Seed planting	2. Seed sprouting
3. Young plant growing	4. Tomatoes forming

Think and do

How can the ripe tomatoes be used to grow other new tomato plants?

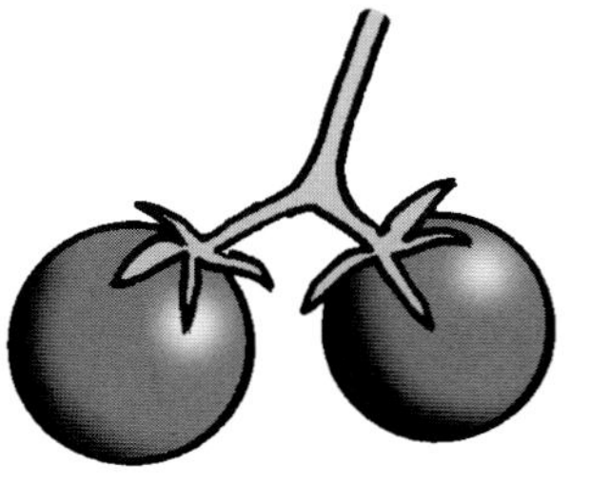

Plants in the cold

Read

A seed that starts to grow is **germinating.**

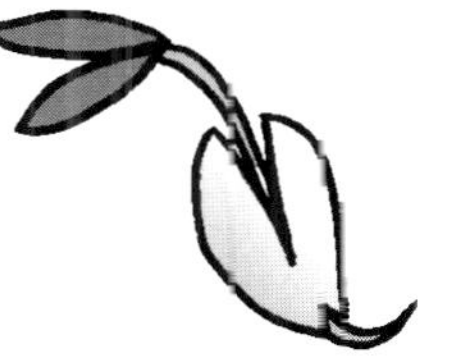

What to do

Which seed will germinate quickest if watered? . . . .

1. Seed in soil in a refrigerator

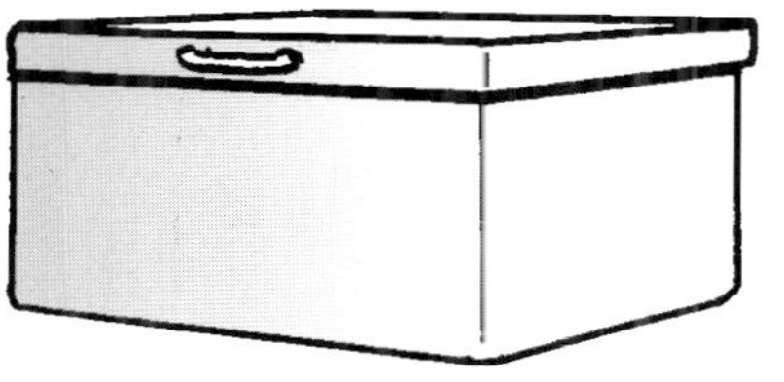

2. Seed in soil in a deep freeze

3. Seed in soil outside in winter

4. Seed in soil in a warm house

You could investigate this.

To start a seed growing, you must have . . . . and . . . .

Think and do

Place a pea or bean seed on some wet cotton wool in the classroom. Leave it for a week or so.

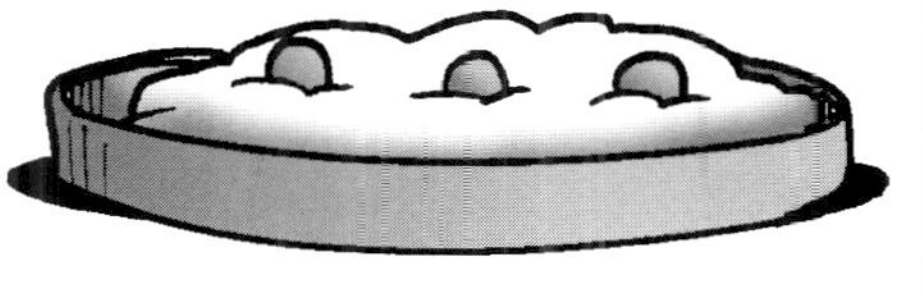

Draw what happens.

What would happen if a similar experiment was done in a dark cupboard?

. . . .

. . . .

Pollination

What to do

Draw lines from the labels to the plant:

Leaves

Petals

Where pollen is found

Where seeds are formed

How do insects and bees help **pollination**?

. . . .

Draw a bee with pollen all over it inside the flower.

Think and do

Pollination is needed because ...

. . . .

Seeds, 2

What to do

Seeds need to be **dispersed** or moved to another place.

Draw lines to show what helps the seeds to be moved.

You can draw more than one line to each plant.

Wind blows

Birds

Squirrels

Humans

Dandelion

Horse chestnut (conker)

Blackberry

Weed seeds

Oak acorns

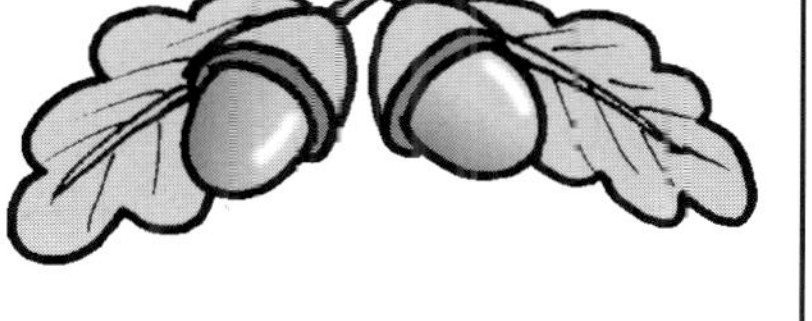

Apples

Different plants

What to do

The words in bold print are mixed up.
Write out what the words should be.

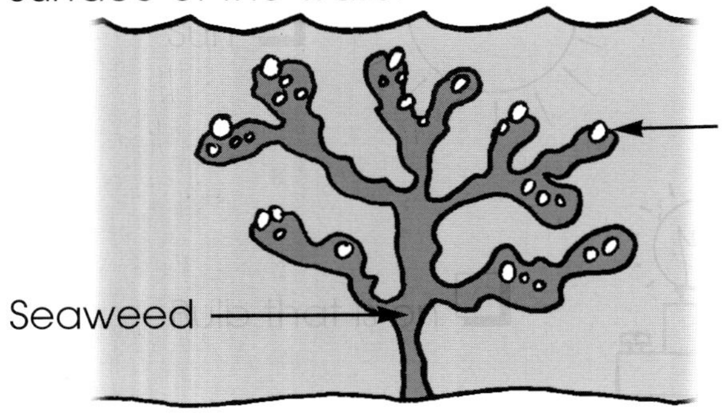

Seaweed uses the air sacks to help it **tolaf** to the surface to get **nus thigl.**

. . . .

. . . .

The ivy clings to the wall using strong **toors**.

. . . .

. . . .

The sweet pea uses a twisting stem to help it **dants pu**.

. . . .

. . . .

Think and do

How do runner beans grow up a stick?

. . . .

. . . .

Guessing animals

What to do

Read these sentences. Guess the living things.

I walk on four legs.
I have a long tail.
I purr when someone strokes me.

I am a **c** . . . .

I grow on a tall plant.
I am a vegetable.
Sometimes people call me a 'runner'.

I am a **b** . . . .

I cannot see very well as I live underground.
People do not like to pick me up as they say I am slippery.
I might live in your garden or in a field.

I am a **w** . . . .

I have a bushy tail.
I love to gather nuts and hide them for the winter.
I sleep most of the winter in a secret hideout.

I am a **s** . . . .

Think and do

Make up a few sentences of your own about a plant or animal.

. . . .

. . . .

. . . .

Test your friends.

What animal?

What to do

A

B

C

D

Use the key to identify the animals.

Has it got six legs?

- Yes → Has it got wings?
  - Yes → Bee
  - No → Ant
- No → Has it got feathers?
  - Yes → Bird
  - No → Has it got fins?
    - Yes → Fish
    - No → Elephant

A is a . . . .

B is a . . . .

C is an . . . .

D is an . . . .

Making keys

What to do

When you make a key you must ask questions that can be answered by **yes** or **no**.
Look at the key for sorting out coins.

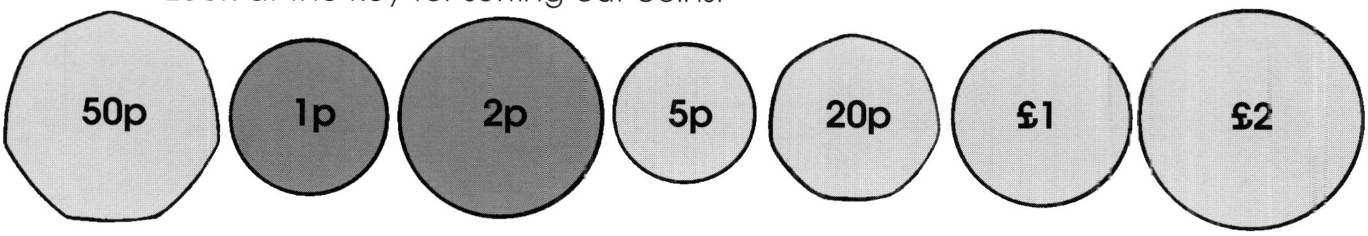

There is no right or wrong way to sort them out. You can make your own yes/no questions.

Work out where the coins should go.

Does the coin have corners?

- **Yes** → Is it smaller than a 10p coin?
  - **Yes** → ☐
  - **No** → ☐
- **No** → Does it have rough edges?
  - **Yes** → Is it the smallest?
    - **Yes** → ☐
    - **No** → Is it the thickest?
      - **Yes** → ☐
      - **No** → ☐
  - **No** → Is it the smallest?
    - **Yes** → ☐
    - **No** → ☐

Write in the box the type of coin.

Think and do

Joe found these leaves in the garden. Make a key to help him sort them.

Holly **Horse chestnut** **Oak** **Apple** **Maple**

Changes in a tree

What to do

A tree changes during the year.
The pictures show how an apple tree changes from spring to autumn.

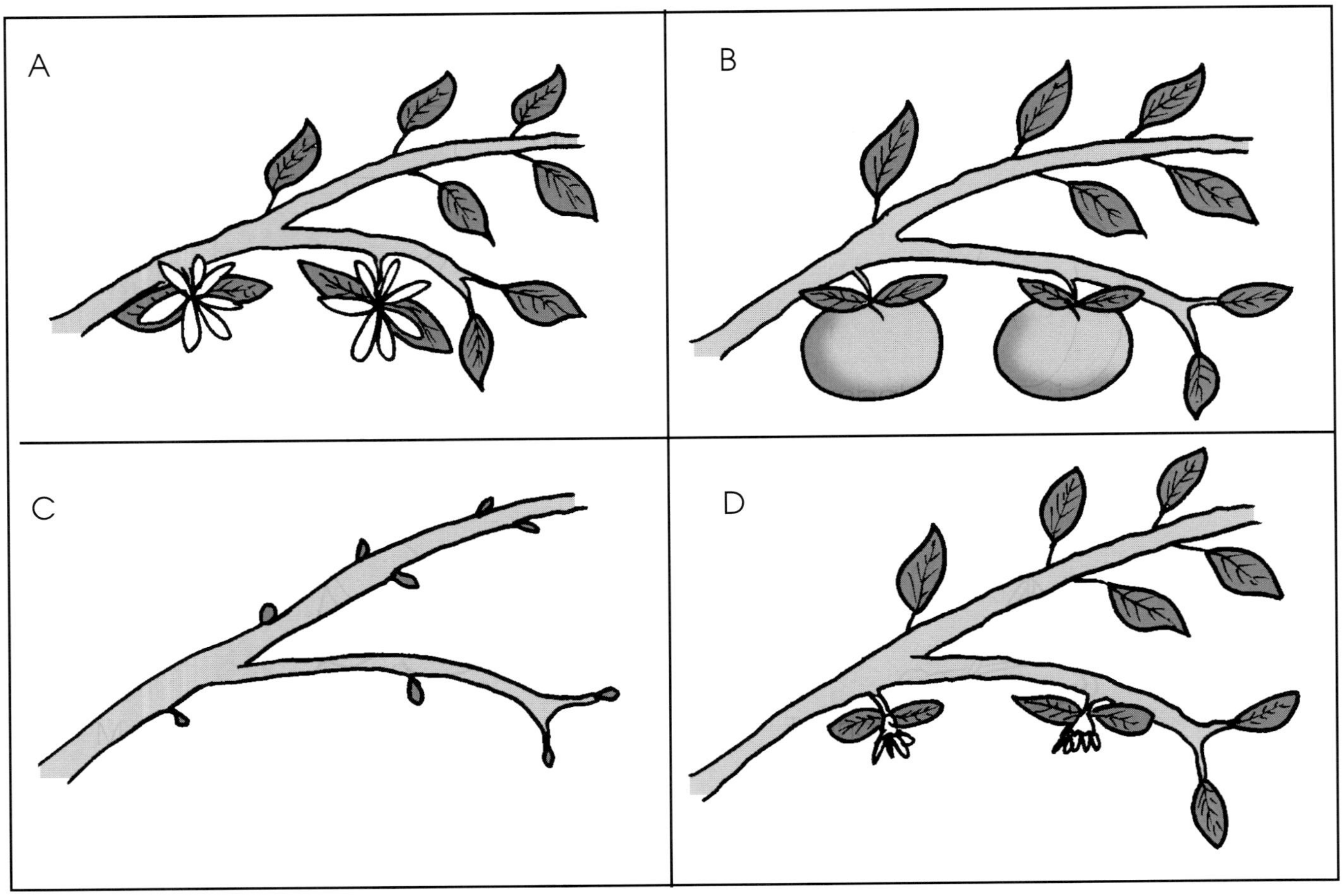

Put the letters in order in the boxes to show how the apple tree changes.
The first is done.

C → ☐ → ☐ → ☐

At what time of year do you see picture C? **S** _ _ _ _ _

Think and do

What season comes after spring? **S** _ _ _ _ _

What season comes after summer? ....

Where do you find the seeds of an apple? ....

Frog cycle

What to do

Cut out the boxes. Put them in order. Start with the eggs.

A

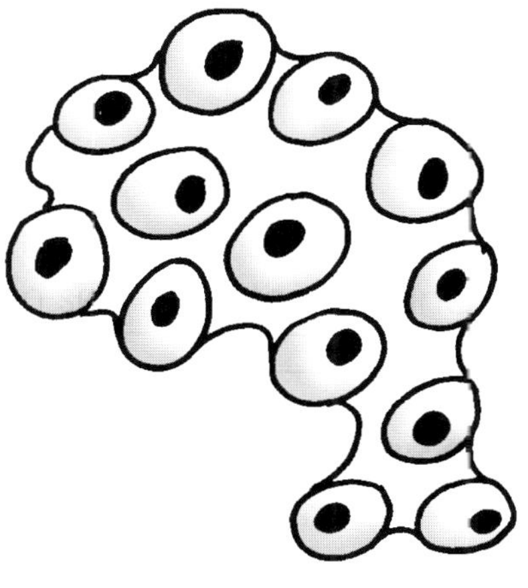

B

C

D

Think and do

What do frogs like to eat?

. . . .

Who lays the eggs – the male or female frog?

. . . .

Mini-beast families

What to do

Kelly was in the woods and she saw some mini-beasts.
She put them into two families.

Family 1

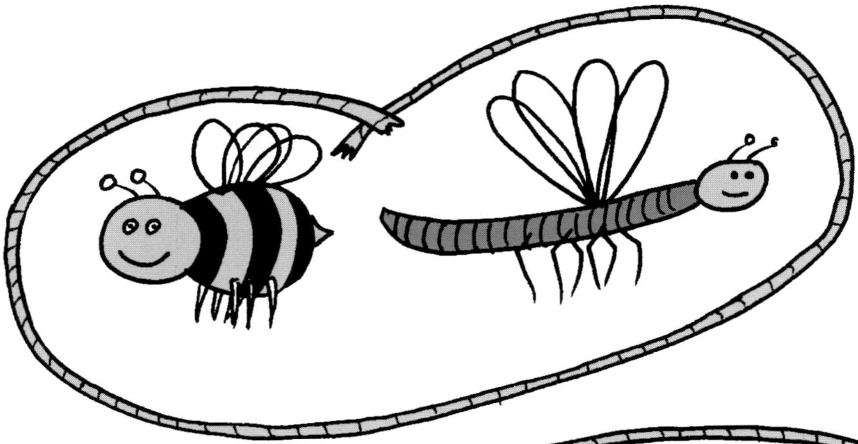

Family 2

How can the mini-beasts in family 1 move ? **f** __ __ and **w** __ __ __

They can do this because they have **w** __ __ __ __ and **l** __ __ __ .

Think and do

How do the mini-beasts in family 2 move?

. . . .

. . . .

Can you name one of the mini-beasts in each family? . . . .

Environment

What to do

The area around the school needs to be kept clean. How can this be done?

Write or draw about ...

trees

rubbish

cars and buses

animal waste

Think and do

How does neglecting these things affect people's health?

. . . .

. . . .

Where do I live?

What to do

Where would you expect me to live?

Draw lines.

Dog

Mouse

Rabbit

Cow

Sparrow

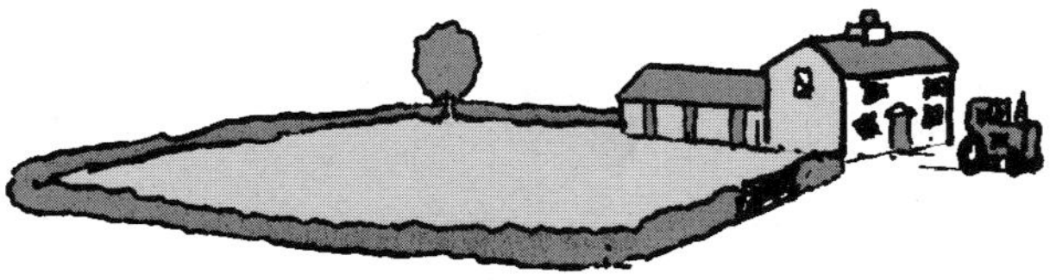

Think and do

Different animals live in different places because ....

....

....

Plants live here

What to do

Draw lines to show where you think the plants grow best.

Heather

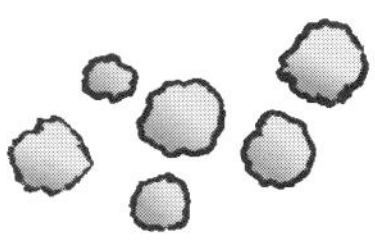

Lichen

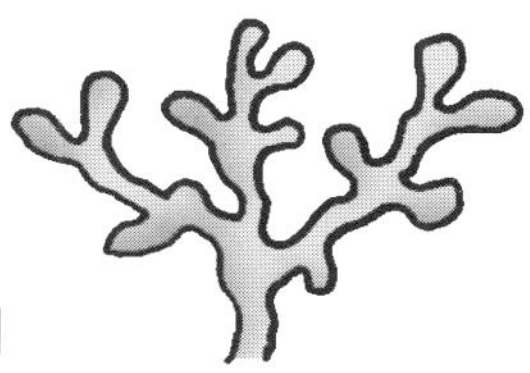

Seaweed

Wheat

House plant

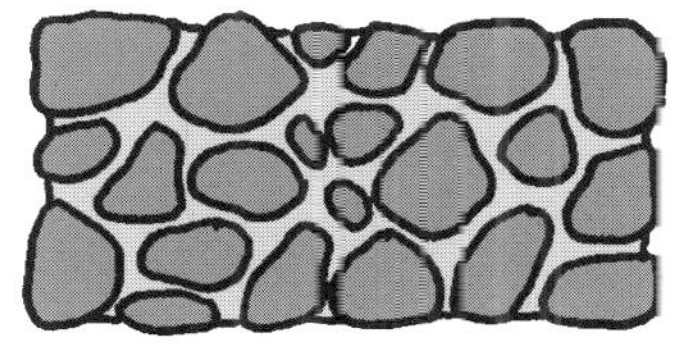

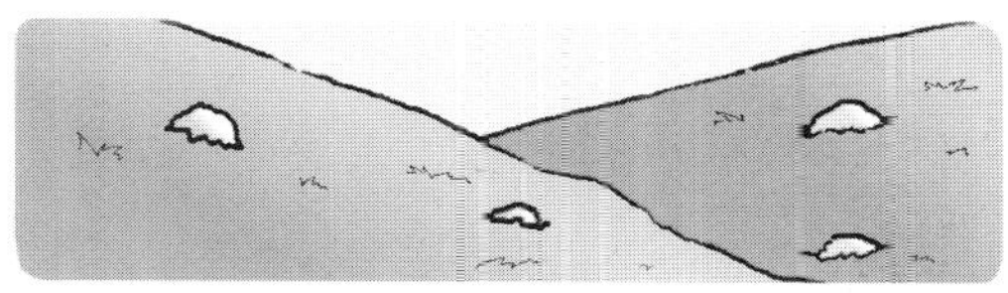

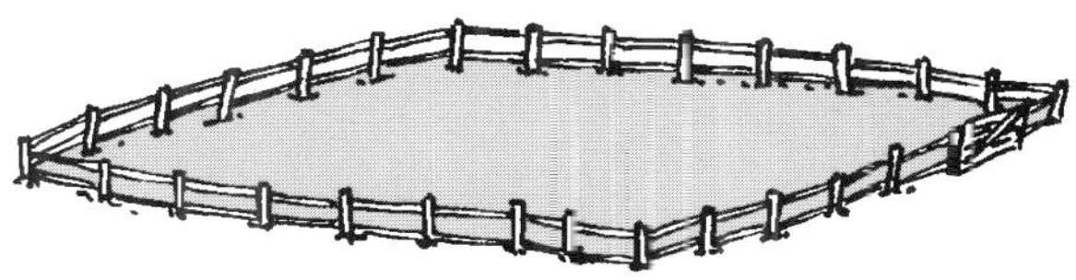

Think and do

Different plants like to live in different places because ....

....

....

Where fish live

What to do

Fish live and swim under water.

Draw lines from the words to the parts of the fish.

Eye

Fins

Mouth

Tail

Gills

Which things help the fish move through the water?

....

....

Fish use **g** _ _ _ _ to breathe under water.

Think and do

Write two ways in which fish are able to live in water.

1. ....

2. ....

Food chains

Read

A cow eats grass → A cow makes milk → I drink milk

This is called a **food chain.** Food chains nearly always start with a green plant.

What to do

A dog eats dog food made from beef. Draw the food chain, starting with grass.

A cat eats 'Catto-fish' cat food. Draw this food chain, starting with plankton.

Jamie likes to eat hamburgers. Draw the food chain.

Think and do

Draw a food chain for a sparrow hawk, starting with a cabbage leaf.

Some animals eat other animals

What to do

Birds eat snails.

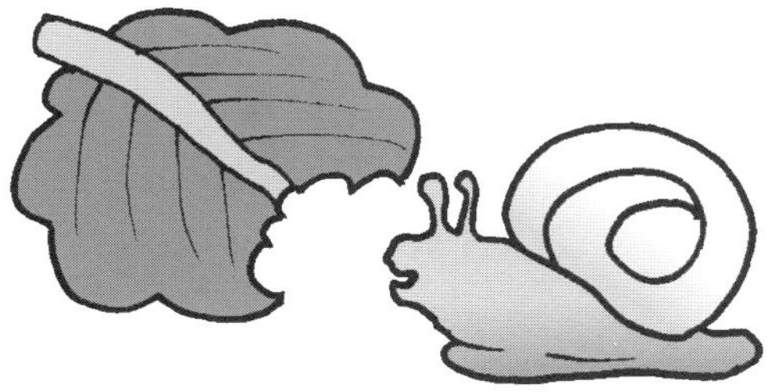

Snails eat leaves.

Slugs eat leaves.

Sparrow hawks eat small birds.

Birds eat slugs.

Give one example of one animal eating another animal.

Name the enemy of the small bird.

Show two food chains, starting with the leaves.

Diseases

What to do

Answer these questions.

I must always wash my hands before eating food because ...

. . . .

I must always wash my hands after going to the toilet because ...

. . . .

I must always use a handkerchief when sneezing because ...

. . . .

Doctors always wash their hands in disinfectant after treating a patient because ...

. . . .

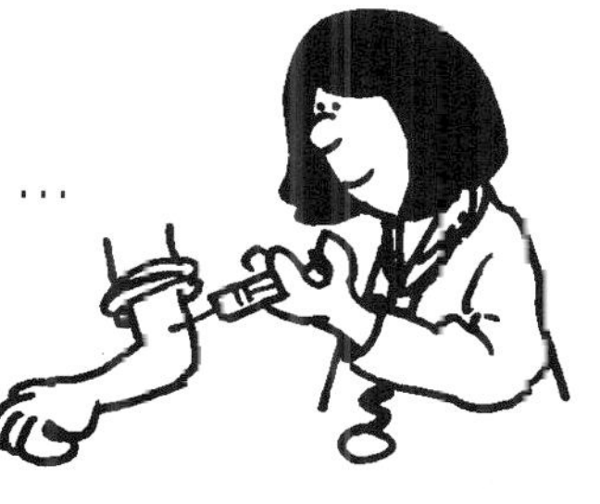

Doctors always use new needles when giving injections because ...

. . . .

Think and do

Complete this rhyme:

Coughs and sneezes

spread **d** . . . .

Micro-organisms

Read

Micro-organisms are living things only seen using a microscope.
Some micro-organisms are helpful. Others are harmful.

Investigate

Dissolve 1 teaspoon of sugar in 150 ml of warm water. Add 1 tablespoon of dried yeast.

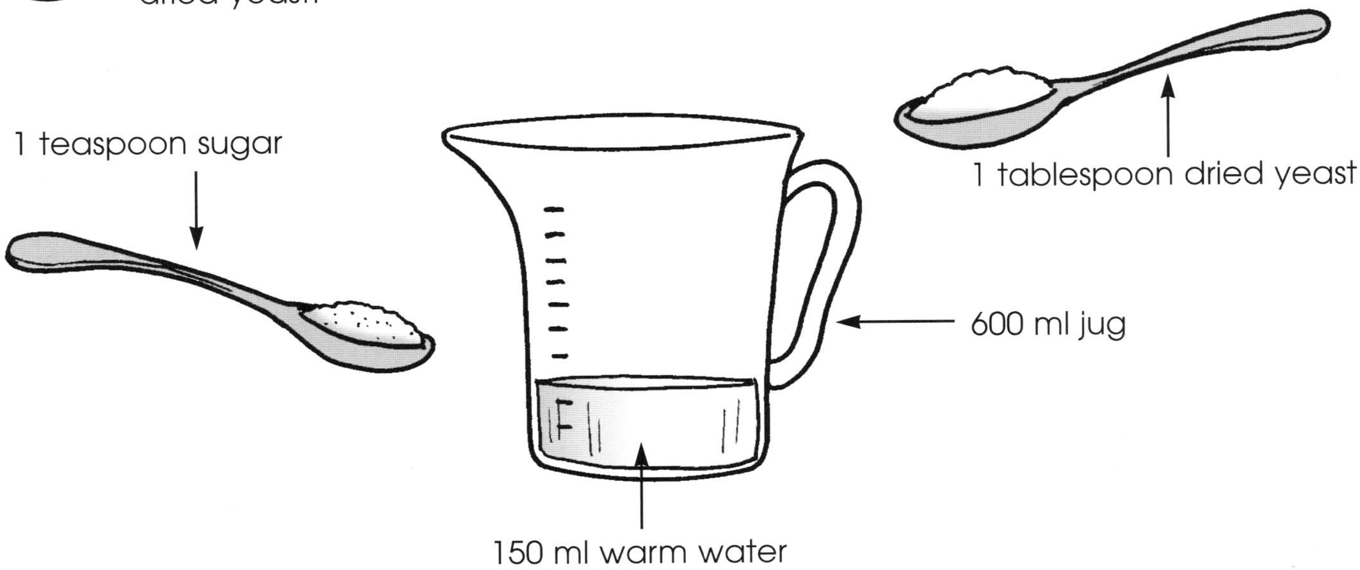

Stir well.

Leave for 15 minutes in a warm place.
Draw what happens.

Cross out the incorrect word:
Yeast is a **helpful/harmful** micro-organism.

You could use your yeast to make bread.

Think and do

Make a list of all the micro-organisms you can think of.
Write if they are helpful or harmful.